Happy Birthday Marshall.

Love,

Grandma

SILKWORMS

Story and photographs by
DENSEY CLYNE

ANGUS & ROBERTSON PUBLISHERS

ANGUS & ROBERTSON PUBLISHERS
London • Sydney • Melbourne

This book is copyright. Apart from any fair dealing for the purposes of private study, research, criticism or review, as permitted under the Copyright Act, no part may be reproduced by any process without written permission. Inquiries should be addressed to the publishers.

First published in Australia
by Angus & Robertson Publishers in 1984
First published in the United Kingdom
by Angus & Robertson (UK) Ltd in 1984

Copyright © Densey Clyne 1984

National Library of Australia
Cataloguing-in-publication data.

Clyne, Densey, 1926-
  Silkworms.

  For children.
  ISBN 0 207 14862 7.

    1. Silkworms — Juvenile literature. I. Title.
    (Series: Young nature series; 8).

638'.2

Typeset in 14 pt Times Roman
Printed in Hong Kong

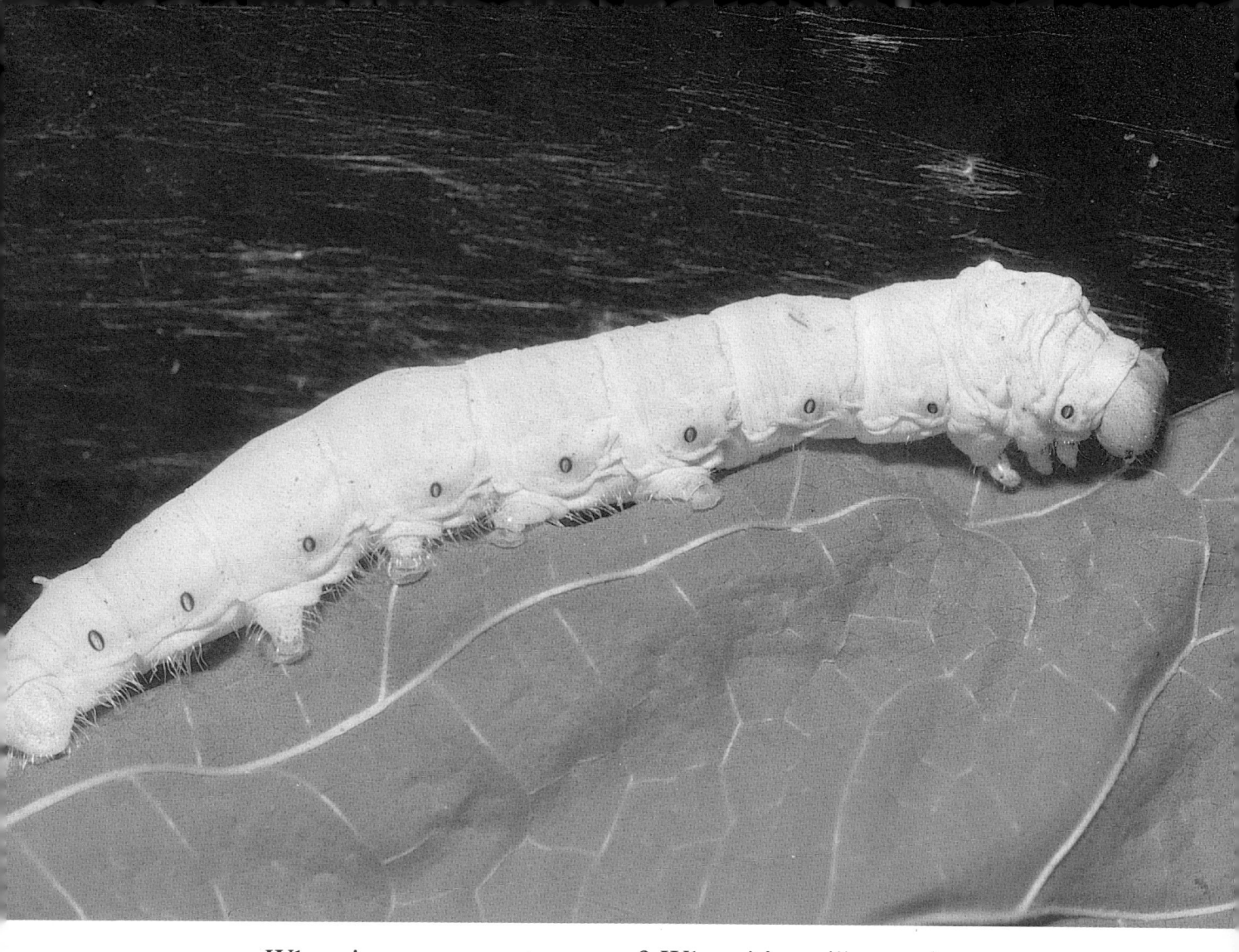

When is a worm *not* a worm? When it's a silkworm!

Do garden worms eat mulberry
leaves, make golden silk,
or grow up to be moths
with creamy-white wings
and feathery feelers?
No, of course not.
But if a silkworm isn't a worm,
what is it?

Yes, it's a caterpillar —
the caterpillar of the
silkworm moth.
You can tell by
looking at it closely.
It has stumpy feet and little
breathing holes along its sides,
as all caterpillars do.

Once upon a time any
little animal that was
long and thin and
wriggly was called
a worm.
Now most of them
have their own special
names. But silkworms
are still called
silkworms.

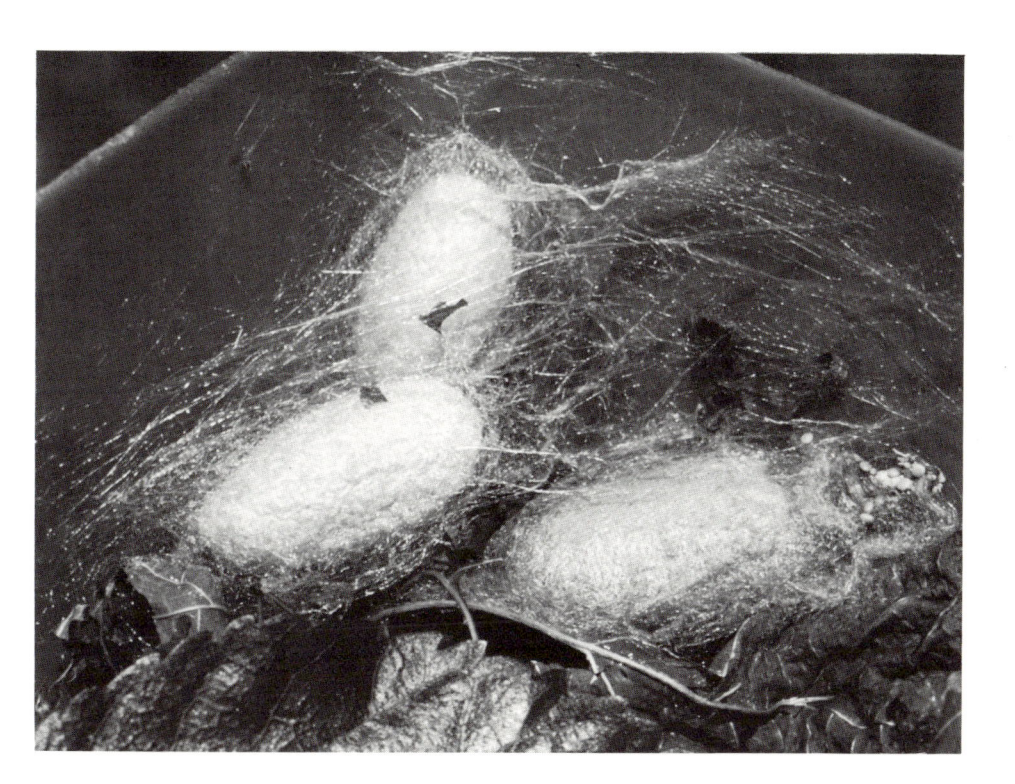

Silkworms are very special caterpillars. Long ago, the people of China found that the silk spun by silkworms could be woven into fine material.
They began to keep silkworms at home and look after them, so that they would be able to harvest the silk.
Now there are no 'wild' silkworms anywhere in the world.

Can you think of
any other insects
that are kept like this?
Yes, we keep bees for their honey
just as the Chinese learned to
keep silkworms for their silk.

Some people started silkworm farms where the silk was taken from the silkworms' cocoons and spun into threads. The threads were dyed different colours. Then they were woven into fabrics, some plain, some with patterns. People from all over the world bought these beautiful silk fabrics and made them into fine clothes.

In China, people became very rich by selling silk to other countries.
But anyone who took silkworms to another country risked having his head chopped off!
Why? Because silk fabric was expensive
and the silkworms that made it were very valuable.

According to a Chinese tale, silkworms were first taken to another country by a princess.
She smuggled them out, hidden in her headdress, as a present for her foreign husband.
Silk is now made in many countries besides China.
And children all over the world keep silkworms.

To keep silkworms you need a mulberry tree in your garden, or a friend with a mulberry tree. Mulberry trees make the garden shady on a hot summer's day. They have delicious berries every year for people to eat. They have plenty of leaves for your silkworms. But imagine all the mulberry trees and leaves needed on a silkworm farm to feed those millions of hungry mouths!

Silkworms are fun to keep, but it must be admitted that compared with some other caterpillars they are rather plain. You could find prettier caterpillars around your garden, or out in the bush.

There are many different
kinds of caterpillar to be seen.
Some are furry. Some are spiky.

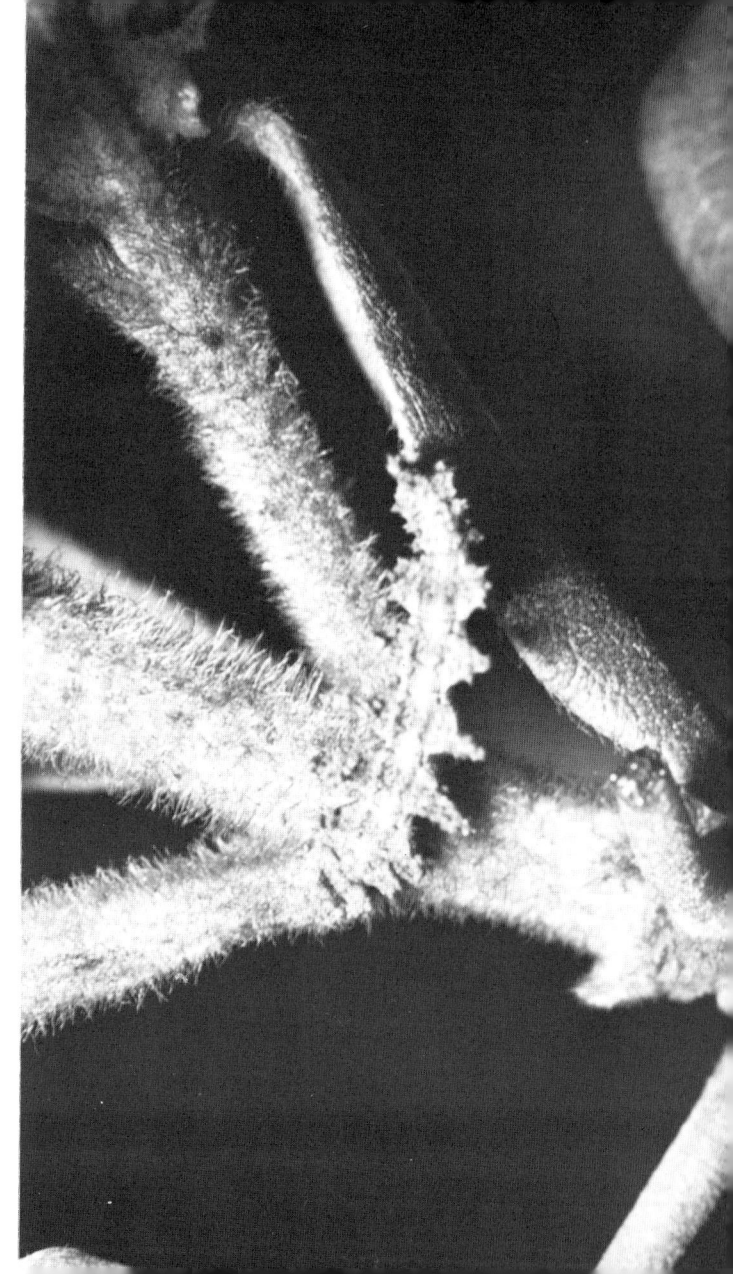

There are caterpillars with unusual shapes.
Some of them don't have any legs in the middle, so when they move they loop along. They are called looper caterpillars.
All these different caterpillars can make silk. The silk thread comes out of a little opening on the caterpillar's lip.
The opening is called a pore.

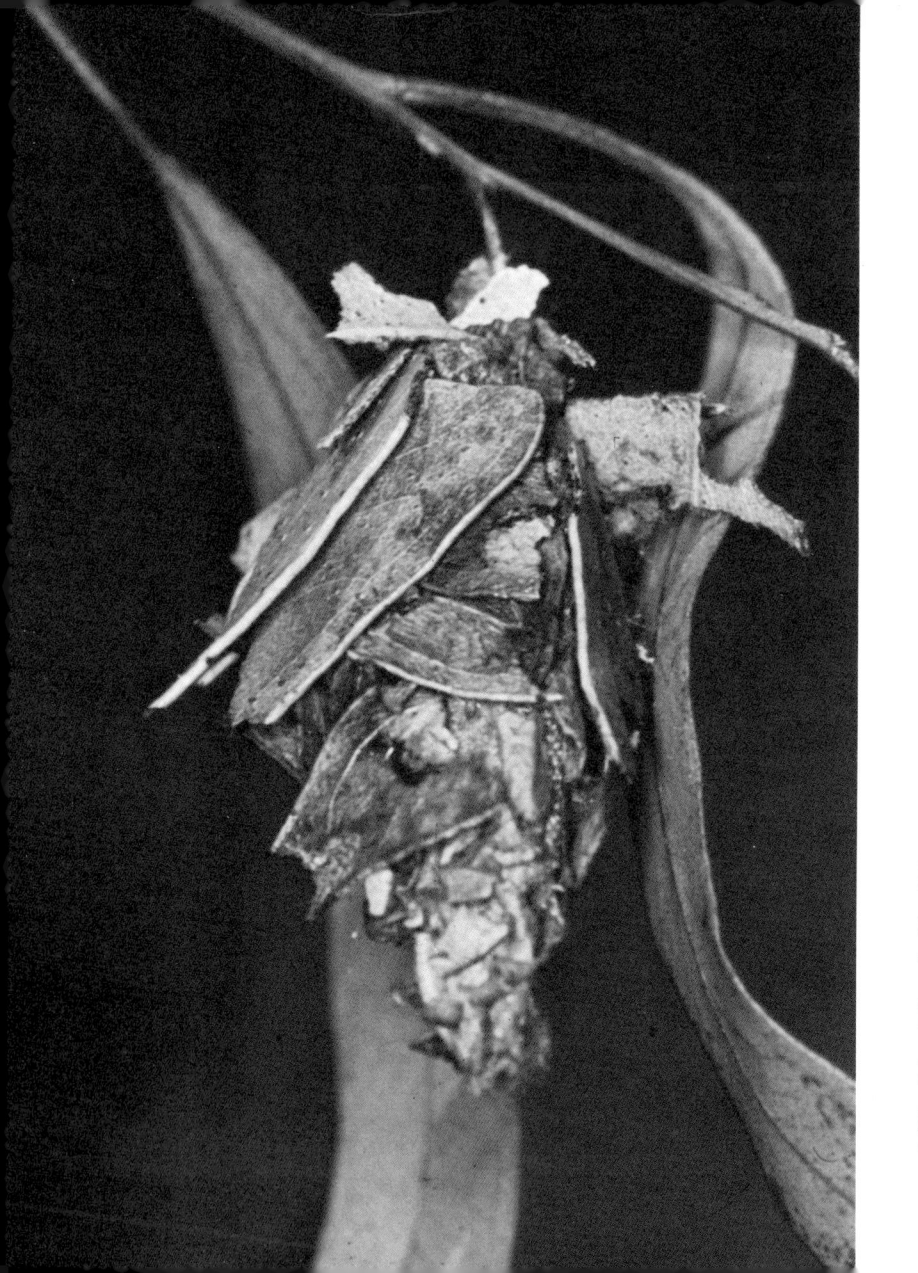

Some caterpillars weave a silk bag
and live in it all the time.
They cover the bag with
bits of leaf. The caterpillar can
poke its head out to feed and
poke its legs out to walk.
The case protects it from enemies.

Silkworms have some wild relatives that can make fine silk, too. Some are Australian, like the hercules moth caterpillar which is much more colourful than its Chinese cousins. But its silk threads are not as long, fine and strong as those of true silkworms.
And you can't keep hercules moth caterpillars as easily as silkworms.

Another wild relative
of the silkworm
is this emperor moth caterpillar.

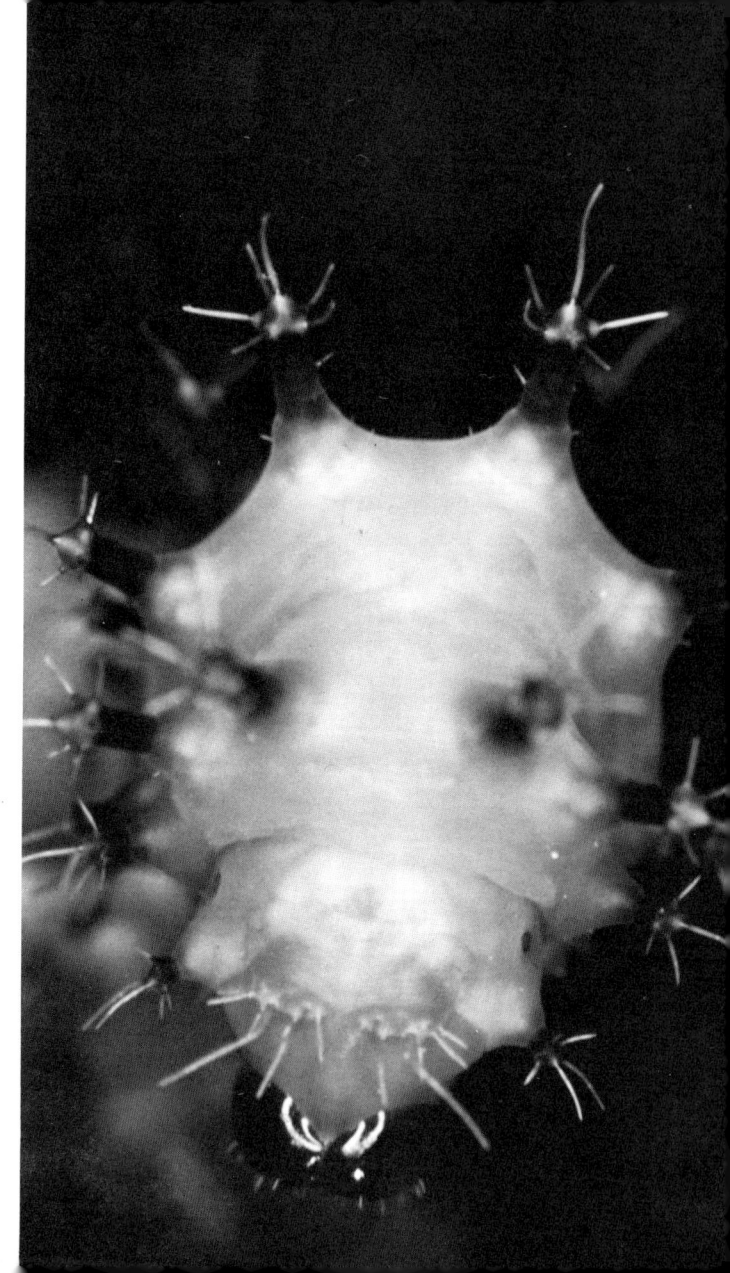

Silkworms sometimes eat lettuce
and other leaves, but their special
food is mulberry leaves.
So you must always make sure
there are plenty of fresh leaves
for them. If you have to get leaves
from a friend's mulberry tree,
you can keep some of them in the
refrigerator, sealed in a plastic bag.
And there's another thing to do.
It's important to clean
the caterpillars' droppings out
of their box every now and then.

As a silkworm grows bigger it has to cast off its old skin when it gets tight.
This is called moulting.

First a new skin forms under the old one.
After a while the silkworm settles down and waits for the old skin to come loose.
Then it just walks forward, leaving it behind!
You can see the old skin behind the newly moulted silkworm in the picture.

A silkworm moults several times during its life.
It grows very quickly.

Soon it's so large that it's ready to turn into a moth.
But before that happens the silkworm must make a safe hiding place.
And that is what a silkworm's silk is mainly used for.
It is used to make a cocoon.

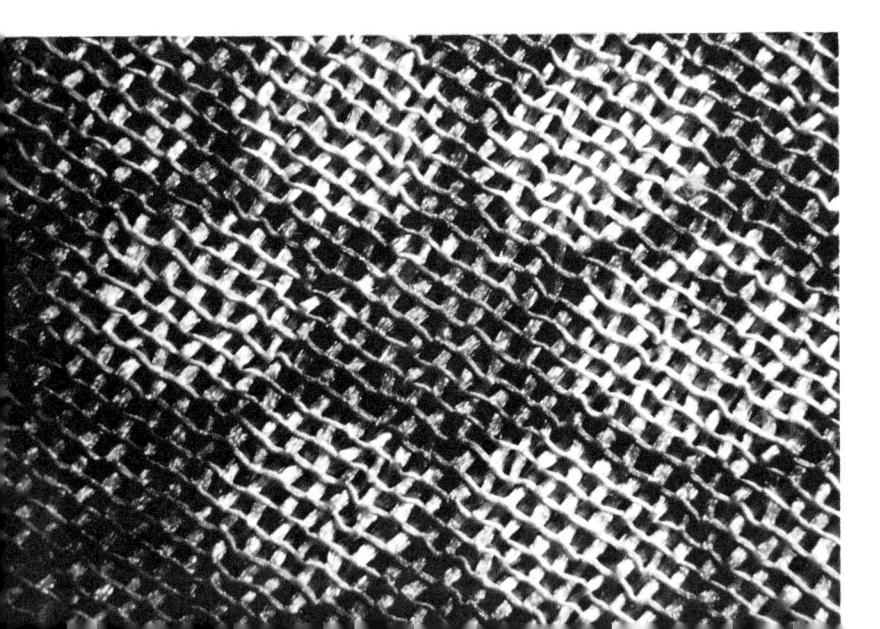

The cocoon looks soft and fluffy.
But it's tough enough to keep the
silkworm warm and dry and safe
while the big change takes place
inside. And, of course, the silk of
the silkworm's cocoon is strong
enough to be woven
into beautiful fabrics.

There are some wild caterpillars that don't make cocoons.
Hawkmoth caterpillars come down from the plants where they feed, and burrow into the ground. They press against the soil to make neat little rooms in which to develop. I had to dig the soil away carefully to photograph this one.

This silkworm's cocoon has been carefully cut open — and look what's inside it!
Not a caterpillar, not a moth, but an in-between stage called a pupa.
'Pupa' means doll, and that is rather what a pupa looks like.
It can't walk or fly or eat or look after itself. That's why
the pupa must lie in such a safe hiding place while it is turning into a moth.

Suddenly, one day, after a few weeks, you'll find your silkworm moths have come out of their yellow cocoons. Almost the first thing a newly emerged moth does is to squirt out a lot of reddish-brown liquid.
Don't worry about this. It's just the waste fluid the moth has to get rid of after being shut in the cocoon for so long.

Silkworm moths don't fly away but they do crawl about. Sometimes they flap their wings so fast that they make quite a loud whirring noise.

Female silkworm moths are a little bit fatter than males when they come out of the cocoon. After a few days the females grow fatter still. That's because they have eggs inside them.

Every year there's a new generation of silkworms to take the place of the old one.
For this to happen the eggs inside the female moth must be fertilised by the male moth. The male and female moths must come together and mate. Then the female lays her eggs.

Wild moths have to travel a long way to find their mates.
But for silkworm moths, all together in their container, it's easy.
I keep my silkworms in a deep dish.
Lots of children keep their silkworms in a cardboard shoe box, and that's all right, too.
If you want to keep the lid on,
you should poke some holes in it to let plenty of fresh air in.

The eggs are laid all over the place, often on the golden silk of the cocoons that the moths came out of.

Silkworms depend on their owners to give them the right kind of food. So you have to keep a watch on the eggs and be ready to feed the caterpillars when they appear.

After a while you can tell that there are caterpillars inside the eggs because of their dark colour.

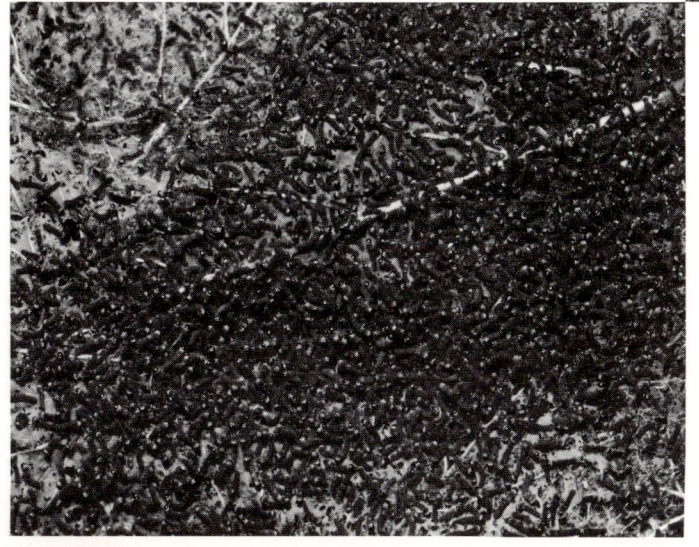

Newly hatched silkworms are dark brown. Then, each time they moult, the new skin is lighter.

Once they've hatched
out of the eggs the caterpillars
will grow and grow. Soon there
will seem to be thousands of them
and you will be wondering where
the next mulberry leaf is coming from
Help! Would anybody
like some nice, healthy silkworms?